Guillermo Bracamonte
Rosalia F. Bracamonte
Lucia V. Bracamonte Gontero

New Technologies based on Nano-, Micro-, Micro-scale control

Guillermo Bracamonte
Rosalia F. Bracamonte
Lucia V. Bracamonte Gontero

New Technologies based on Nano-, Micro-, Micro-scale control

For the release of drugs, nutraceuticals, and non-classical light.

ScienciaScripts

Cover image: www.ingimage.com

This book is a translation from the original published under ISBN 978-620-0-01938-7.

Publisher:
Sciencia Scripts
is a trademark of
Dodo Books Indian Ocean Ltd. and OmniScriptum S.R.L publishing group

120 High Road, East Finchley, London, N2 9ED, United Kingdom
Str. Armeneasca 28/1, office 1, Chisinau MD-2012, Republic of Moldova, Europe
Managing Directors: Ieva Konstantinova, Victoria Ursu
info@omniscriptum.com

Printed at: see last page
ISBN: 978-620-8-57796-4

NEW TECHNOLOGIES BASED NANO-, MICRO-SCALE CONTROL FOR DRUG DELIVERY, NUTRACEUTICALS AND NON-CLASSICAL LIGHT

VERSION EN ESPAÑOL

NEW TECHNOLOGIES BASED ON THE NANO-, MICRO- SCALE CONTROL FOR PHARMACEUTICS, NUTRACEUTICS, AND NON-CLASSICAL LIGHT DELIVERY

ROSALIA F. BRACAMONTE, SPECIALIST. SCIENCE TECHNICIAN. [6], LUCIA V. BRACAMONTE GONTERO, SPECIALIST. CHEMICAL TECHNICIAN. [6], A. GUILLERMO BRACAMONTE, DR. IN CHEMICAL SCIENCES [1], [2], [3],

*[4], 5], [6], [7]**

[1] *INSTITUTO DE INVESTIGACIONES EN FÍSICO QUÍMICA DE CÓRDOBA (INFIQC), DEPARTAMENTO DE QUÍMICA ORGÁNICA, FACULTAD DE CIENCIAS QUÍMICAS, UNIVERSIDAD NACIONAL DE CÓRDOBA (UNC). CIUDAD UNIVERSITARIA, 5000 CÓRDOBA, ARGENTINA.*

[2] *DEPARTEMENT DE CHIMIE AND CENTRE D'OPTIQUE, PHOTONIQUE ET LASER (COPL), UNIVERSITÉ LAVAL, QUÉBEC (QC), CANADA, G1V 0A6.*

[3] *DEPARTMENT OF CHEMISTRY, UNIVERSITY OF VICTORIA (UVIC), VANCOUVER ISLAND, V8W 2Y2 BRITISH COLUMBIA (BC), CANADA.*

[4] ***INSTITUT FUR ORGANISCHE CHEMIE, UNIVERSITAT REGENSBURG 93040, REGENSBURG, GERMANY.***

[5] ***UNIVERSITY OF AKRON, INSTITUTE OF POLYMER SCIENCE AND ENGINEERING; AND NASA ASTROBIOLOGY INSTITUTE, OHIO, UNITED STATES.***

[6] ***(I) ESCUELA TÉCNICA ENET N°1, GRAL. SAVIO (IPET Nº 266 GRAL. SAVIO); AND (II) REMEDIOS ESCALADA DE S. MARTÍN. CÓRDOBA, ARGENTINA; III) SCHOOL COMMISSION "DES DECOUVERTES", QUEBEC; AND "BRITISH COLUMBIA EDUCATION SYSTEM", BC, CANADA.***

[7] ***MEMBER OF THE TECHNOLOGY LIAISON NETWORK OF CONICET-CCT (CENTRO CIENTÍFICO TECNOLÓGICO), CÓRDOBA, ARGENTINA.***

UVictoria ULaval

URegensburg UAkron

INDEX

PREFACE

This research work develops the topic of new technologies based on the control of matter and its composition at the Nano-, and Micro-scale for the generation of Nanotechnology, Micro-devices, Chips, and Instrumentation of reduced dimensions. In particular, it focuses on the design of ideas and new strategies for the release of drugs, bio-drugs and nutraceuticals. That is to say, on different chemical species of interest in Life Sciences with interests in health, nutrition and sports performance improvement. It also addresses the subject of diagnosis and treatments where a high level of instrumentation and related technology is involved and where new Nano-, Micro-materials can be part of a new technique or methodology. In other words, recent advances in progress in this respect are shown and analysed, as well as concrete and real incorporations are compared and mentioned which are already on the almost daily market. Similarly, the aforementioned concept of release of a chemical species associated with a certain property has been extended to the release of non-classical light. This so-called non-classical light relates to new strategies in the generation of light that involve manipulations of wavelengths, energies, and colourings with various applications, just as in the classical concept. Examples of nanoemitters and related technologies such as LEDs (Light Emitting Devices), lasers, and lamps are shown, where they are applied from studies in cells to current lighting technologies. In addition, knowledge and awareness of the need for the use nanoemission technology is being expanded. light in new health treatments as well as for the development of a good life in general. It is thus a compendium of sample and analysis information for the proposal of new technologies in the context of education, science and development. In this sense it also shows recent experimental results in development and published by the authors on new nanotechnologies and applications. In this way, information has been collected from different Universities and Education systems in different countries and regions of the planet in a face-to-face manner; as well as many others developed in virtual activities fulfilling tasks related to congresses, courses, distance education, editorial projects and collaborative works.

SUMMARY

In this treatise on new emerging technologies in the field Life Sciences, the construction of knowledge from Nano-, Micro-scale control to the proposal and development of Nanotechnology, Microtechnology, Microdevices, Chips, and instrumentation of reduced dimensions is approached in a simple way. The control of tiny particles is approached by chemical reactions in colloidal dispersion as well as by other dry technologies using lasers and methods for the modification of surfaces and substrates. These Nano-, Microstructures can have different applications. In this sense, the present research work has developed the design concept with perspectives towards the preparation of new nanopharmaceuticals and nutraceuticals for the controlled release of their active ingredients. Also for the development of early diagnostics and application of new high precision medical treatments. Many of these concepts are currently under study at the level of fundamental science, but many of them transcend the market and are already incorporated in technologies and treatments to which society has access without knowing it. Furthermore, from the nano-, micro- and micro-scale, non-classical light release can be generated for different reasons. Nanoemitters with high light intensity can mark cells, tissues as well as act as nanoplatforms where molecules of biological interest can be detected. Similarly, these tiny architectures can participate in the generation of light in new systems such as LEDs (Light Emitting Devices). In this sense they can act as release of light for illumination of optical systems such as microscopes as well as irradiating cells and tissues for various applications related to Bioimaging and Biodetection.In this compendium of knowledge, all these topics are dealt with in a multidisciplinary approach with emphasis on Chemistry, Nanochemistry, Nano-optics, Physics, Photophysics, and Photonics developed with Biophotonics and Nanomedicine perspectives. It is there, where in the same structure of reduced dimensions that can take the form of a powder, a colloidal dispersion, a cream, and a dressing can contain Nano-, Microstructures with functionalities of release of a certain drug, Nutraceutical product, and also release of non-classical light detected by means of sophisticated Optical systems of high precision such as Microscopes, Tomographs, and scanners. For the presentation of these topics we have analysed recent publications as well as pioneering articles in each of the scientific disciplines, which has allowed us to present a brief summary or review of the topics. It should also be noted that this treatise has been developed in the context of technological linkage, Science Education, and Continuing Education, in various scientific disciplines and transversal areas of knowledge which are in collaborative projects with different International Academic Institutions. , at the end of this compendium, advances in multifunctional nanotechnologies incorporating multiple functional materials for their controlled release by means of remotely controlled systems or mechanisms are briefly presented. Thus, it discusses and concludes on advances from fundamental science to knowledge transfer in the field of nanotechnology. Life Sciences market where we are all in search of new strategies for the improvement of the quality of life in general.

INTRODUCTION

1. Introduction: concepts from an active principle of a molecule to Nanochemistry in Life Sciences.

This introduction deals with the basis and fundamentals on which knowledge is built from the design and synthesis of one molecule to assembly of several molecules as they occur in polymeric reactions. Continuing with the control in 3 dimensions it is possible to assemble molecules and atoms in higher dimensions to generate Nano-, Micro- and Micro-materials with certain functions. In this context it is mentioned that the average unit of the nanometre relates length dimensions in millionths a millimetre smaller than a millimetre available in school rulers. Thus, the control of the nanoscale is at infinitesimally small sizes but well above the molecular level. To understand this, it is necessary to know that molecules with average dimensions such as amino acids in proteins can agglomerate together from 100-1000 to form 1 nanometre (nm). It is in these dimensions that the development of this compendium Nanotechnology and Life Science applications has taken place.In particular, it develops how nanoscale control can be used for the design and synthesis of new nanopharmaceuticals and nutraceuticals. Various molecules with active ingredients can have positive effects on health; however, one of the biggest challenges in pharmacy is to the route of administration, incorporation and sensing of the tissue or organ so that it initiates its desired action at the appropriate site. Thus, not all of challenges have been addressed and many of them have been partially addressed. A given drug may be degraded in the different media through which it circulates until it reaches the specific site where it must act based on specific interactions with target receptors.

Similarly, the incorporation of Nutraceuticals can be contemplated, being a group of synthetic and natural molecules; Biostructures, Nutrients, antioxidants, vitamins and various chemical structures which contribute to nutrition for the good conformation and maintenance of the organism with full health. The main objective of Nutraceuticals is the protection of health through the controlled intake of specific nutrients, which are incorporated in advance for the body's disposition in the anabolic process of construction and maintenance of vital functions through various biological mechanisms. In this way, they can act as preventive agents in the development of future ailments. In this sense, Nutraceutical agents can be considered as pharmaceuticals as well. And in that sense the release of pharmaceuticals and Nutraceuticals in a controlled manner, and activated by mechanisms which are considered to intelligent based on a specific response are also some of the major challenges to be addressed from the point of view of drug design and application in a given treatment.

In this sense, nanoscale control can bring many improvements in the absorption capacity of drugs and nutrients. This is related to the interactions of the molecules involved with the medium in which they are absorbed. The pathway and delivery to the target organ and/or membrane. It can also be as a structure of minute dimensions that provides controlled concentrations of specific molecules for local application. And in this context local application relates to the level of cells and organelles as well. This has the implication that a drug can be dosed in a controlled manner at an active site that needs only the treatment. In this way the treatment increases its specificity and decreases possible unwanted adverse effects.[1] , the micro-scale can provide in materials, the design of larger surfaces and volumes involved which can contain Nanopharmaceuticals and Nutraceuticals. In other words, the control of matter in the manufacturing process of a material can provide new forms of administration. These can be contemplated in the design of hydrogels, capsules, tablets and new application strategies based on tissue contact such as dressings, devices, and chips.[(2)] These new routes of administration can be found conceptually in some pharmaceutical solutions implicit in product design today. But, in this regard, there are many advances in new technologies based on fundamental science where development in this direction is taking place with a variety of applications and interests.

In addition, new technologies include light-emitting systems, many of which can be considered as non-classical light delivery systems such as lasers.[(3)] The control of tiny particles is proposed by chemical reactions in colloidal dispersion as well as by other dry technologies through the modification of surfaces and substrates. These Nano-, Microstructures can have different applications. In this sense, the present research work has developed the concept of design with perspectives towards the preparation of new nanopharmaceuticals, nutraceuticals and controlled release systems of non-classical light in a similar way.[4] their therapeutic properties and assistance in various surgical procedures.[5] Also for the development of early diagnosis and application of new high-precision medical treatments. Many of these concepts are currently under study at the level of fundamental science, but many of them transcend the market and are already incorporated in technologies and treatments to which society has access without knowing it. Moreover, at the nano-, micro- and micro-scale, non-classical light release can be generated for different reasons.[6] Nanoemitters with high light intensity can mark cells and tissues as well as act as nanoplatforms where molecules of biological interest can be detected. Similarly, these tiny architectures can participate in the generation of light in new systems such as LEDs (Light Emitting Devices).[7] In this sense, they can act as light delivery systems for the illumination of optical systems such as microscopes as well as irradiate cells and tissues for various applications related to bioimaging and biodetection. Thus, this compendium of knowledge deals with all these topics in a multidisciplinary approach with emphasis on Chemistry, Nanochemistry, Nano-optics, Physics, Photophysics, and Photonics developed with perspectives of Biophotonics and Nanomedicine.[(8)] It is there, where in the same structure of reduced dimensions that can take the form of a powder, a colloidal dispersion, a cream, and a dressing may contain Nano-, Microstructures with drug delivery functionalities, Nutraceuticals, and also non-classical light release detected by sophisticated high-precision Optical systems such as Microscopes, Tomographs, and scanners[(9).]

OBJECTIVES

2. Objectives

• The general objective is to develop and deal with the subject of Nanochemistry control for the generation of new structures at the nanoscale and microscale to contain and release drugs and nutraceuticals in an improved way in various aspects required in the field of Life Sciences.

• Furthermore, and similarly, the need to generate new systems of controlled light release from lower dimensions such as the nano-, and micro-scale is analysed. Thus, non-classical light is presented from a molecular point of view and how to stimulate light phenomena in confined chemical structures.

• On the other hand, after having generated the necessary knowledge for the control of the Nanoscale, analyse different developments focused on the functionalisation of materials towards a larger scale such as devices, portables, and Chips. In these can be contained the different Nanotechnological developments confined in volumes or placed on surfaces to design and create physical and chemical properties with specific functions. The functions in particular focus on the three main themes of pharmaceutical delivery, nutraceuticals and non-classical light.

• It is also proposed to develop the design and manufacture of nanopharmaceuticals. These have particular properties and advantages, many of which are being studied at the level of fundamental science and show interesting perspectives.

• The need to develop new systems for the controlled release of nutraceuticals, which can have a wide variety of objectives, will be studied. In this sense, the development of nanoneutraceuticals will be studied from the point of view of direct interaction with target cells in need of a specific nutrient.

• The development of Nanoemitters based on the control of physical andchemical phenomena is developing towards the controlled release of non-classical light. This The concept also aims to analyse the current state of art development on a fundamental level as well as applied in new technologies already on the market such as LEDs and lasers.

• Finally, a multifunctional nanoarchitecture design is presented based on a heart or core with nanoscale dimensions of a biocompatible material on which polymeric multi-coatings are added. The concept is described, analysed and discussed about properties which can be considered as Nanoemitters, Nanoconductors, Nanopharmaceuticals and Nutraceuticals. In this context, results and advances obtained in Chemistry and Optics Laboratories are shown, highlighting recent results and future perspectives.

• Then, the design and control of materials towards a larger scale is considered, contemplating from the micro-scale to the miniaturisation of instrumentation and functional devices. In this context, the approach of prototypes is analysed from the concept to the actual transfer of the creation of the material with the desired properties.

MATERIALS AND METHODS

3.1 Materials

As for the materials, they can be classified into three main sources: i) Bibliographic sources, media, and Electronic Network; ii) academic sources based on study stays, visits, face-to-face work, remote work, collaborative work, and through related courses and memberships. It should also be noted that this treaty has been developed in the context of technology linkage, Science Education, and Lifelong Learning, in various scientific disciplines and cross-cutting areas of knowledge which are in collaborative projects with different International Academic Institutions. (iii) materials related to chemical reagents for the conduct of chemical reactions to obtain molecules, modified molecules and Nanostructures[10]; (iv) materials related to chemical reagents for the conduct of chemical reactions to obtain molecules, modified molecules and Nanostructures.

3.2 Methods

The bibliography was duly compiled, organised and summarised in most cases. For the analysis of the literature used, a reading of the literature has been carried out, extracting the main idea without emphasising an extensive description, but instead extracting the one related to the desired communication in the search to fulfil the stated objective. The detailed description can be found in the original publications where applicable. In terms of experimental development, different techniques and methods were used for the control of nanochemistry in colloidal dispersions.

From the experimental results, various adaptations and modifications of known methodologies were developed and new ones were generated as appropriate.

The synthesis of metal nanoparticles was performed by reduction , such as the classical Turkevich method[11]. The synthesis of silver nanoparticles was carried out in a similar way. By varying the amounts of reactants it was possible to obtain nanoparticles with dimensions between 10-200 nm.

Furthermore, to obtain hybrid nanoparticles, these nanoparticles were coated with a polymer, silica (-SiO_2-), and a biodegradable polymeric coating by the Stöber method;[12,] and spontaneous emulsification by solvent diffusion, respectively according to the application. In this way, the nanoparticles, besides being stabilised in aqueous media, can covalently bind functional groups, fluorophores, supramolecular systems, proteins, antibodies, etc., by means of organic reactions.Biodegradable polymers are copolymers formed from the combination of monomer pairs such as lactic, gallic and malic acids, among others. Their synthesis consists of a spontaneous emulsification method by solvent diffusion[13].

The most important parameters to take into account in the synthesis were the emulsifier (PVA, molecular weight 20000) and its R= copolymer/emulsifier ratio; and also the ratio between the solvents used (Rs= H_20, Acetone, Ethanol). For the synthesis of Supramolecular Polymeric Nanoparticles, according to the design concentrations variable of cyclodextrins, and modified cyclodextrins[14] were added to the polymeric nanostructure by simple cross-linked incorporation or via covalent bond formation[(15).]

RESULTS. DATA AND ANALYSIS

4.1 Design y synthesis of Nanoparticles, Nanoplatforms, and Microdevices with applications in new technologies.

Drugs are widely used in different formulations and dosages depending on the disease. The development of biodegradable nanostructures in biomedicine has solved problems with the use of conventional microparticulate drugs[16, 17]. Their bioavailability can be increased, allowing for improved dosage, prolonging the therapeutic effect and reducing adverse reactions. Its specificity is increased through the incorporation of biomolecules and antibodies, which also allows the local concentration of the drug to be increased. In addition, complementary treatments can be coupled for greater effectiveness. The use of these nanoparticles allows for increased solubility due to their higher area/volume ratio and thus higher bioavailability and lower toxicity[18].These nanostructures, varying in size between 10-500 nm, can be can be obtained through the use of various chemical structures. Supramolecular chemistry provides diverse structures with the ability to form host-receptor inclusion complexes and self-assembled arrangements based on non-covalent interactions.[19] On the other hand, polymer chemistry allows structures with diverse properties to be obtained depending on their constituent monomers. From the combination of both areas of chemistry, modified polymers can be obtained with supramolecular systems which possess properties resulting from the combination of each of their components. In addition, hybrid organic materials in the presence of inorganic nanoparticles give them metallic, magnetic, conductive, plasmonic, etc. properties.[20] New design concepts are emerging from these materials, such as nanoparticles with a metallic core coated with molecular spacers,[21] polymeric,[22] containing drugs of pharmaceutical interest, to which biomolecules, antibodies, etc. are added in order give them specificity[23].

On the other hand, nanoparticles formed by modified liposomes, short-chain polymers, micelles and organised hybrid systems composed of metals have been developed with certain physical and chemical properties that allow a drug to be released with high efficiency and low toxicity[24,25]. Finally, it is important to mention that in addition to controlled release applications, these structures have diverse applications such as biological event detection, cell detection and tracking, diagnostic imaging and new therapies in nanomedicine that provide great future prospects in the pharmaceutical market.[26]

4.2 Nanopharmaceuticals and controlled drug release.

Nanopharmaceuticals are related to the incorporation of active ingredients from molecules with confined pharmaceutical properties. at the nanoscale in order to improve some aspect required in their application or treatment.[27] Drugs which are conceived at the molecular level need protection from external agents in the course of their incorporation into the organism. Thus, it is necessary to create strategies for their administration that include formulations with protective agents. Thus, drug-loaded nanoparticles can be proposed in many cases as vehicle and controlled-release systems.[(28)]

In addition, nanoscale control can provide functional structures which intrinsically act as nanodrugs with an active principle or site[29], which elicit a desired response in the organism.In this sense colloidal systems exist widely in the pharmaceutical market through specific formulations with the incorporation of Nanoparticles being a trend in the next generation of drugs called Nanodrugs.[(30)] This concept can currently be seen in progress by offering new products with improved properties and/or towards new solutions to current challenges such as in the area of Nanovaccines. [31]

4.2.1 Biodegradable polymeric nanoparticles

There are several nanoparticles synthesised based on different monomers with different properties, such as gallic acid, lactic acid, malic acid, etc.[32] These nanostructures allow the incorporation of pharmaceutical drugs, either by simple encapsulation or by means of the chemical modification of organic functional groups.[33] The incorporated drugs can then be released in a controlled manner based on the hydrolysis times of their covalent bonds depending on the conditions of the medium or external stimuli to trigger the process. Such processes are part of the challenge in the design of smart materials.[34] In this area, our laboratory has synthesised polymeric nanoparticles, formed by co-polymers of lactic and malic acid, which possess non-toxic properties due to the incorporation of a monomer that is easily biodegraded by the organism.[35] The nanoparticles obtained have dimensions between (200-300) nm (**Figure 1**), and fluorescent molecules have been incorporated which allowed controlled release studies to be carried out. This synthesis of biodegradable polymeric nanoparticles has subsequently been used to synthesise hybrid organic-inorganic materials.

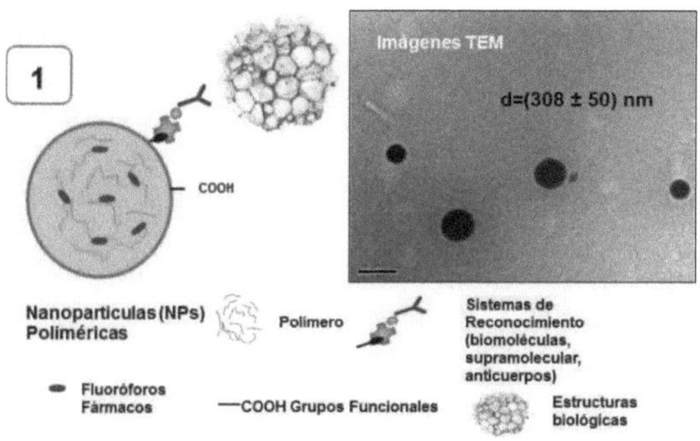

Schematic of Biodegradable Nanoparticles with incorporation of Transmission Electron Microscopy (TEM) image of Poly-L-lactic-, Malic Acid Nanoparticles. Photos obtained at Laval University, Quebec, Canada. Reprinted with permission from A. G. Bracamonte et al. 2024.

4.2.2 Nanoassemblies, Supramolecular Nanoparticles, and Organised Systems

Nanoparticles based on supramolecular interactions have shown great development in recent years.[36] The most widely used supramolecular systems for the interaction and formation of nanoaggregates are macrocycles. These structures are cyclic oligomers with a conical or cylindrical shape, which possess a nanocavity that allows the incorporation of organic drugs,, and other chemical species[37]. In addition, these structures possess a synthetic versatility that allows them to be chemically modified and incorporated into flexible polymeric chains, which form through non-specific interactions Nanoaggregates. In this way, the incorporation of organic drugs can be achieved by the formation of inclusion complexes with the macrocycles or by incorporation into the polymeric network[(38).] It is important to note that the first therapy based on macrocycle nanoaggregates applied in humans for targeted release of RNA (SiRN A) was developed with self-assembled cyclodextrins[.(39)] Preliminary results obtained in our laboratory using biodegradable nanoparticles combined with covalently modified cyclodextrins show the formation of nanoparticles between (300-600) nm (**Figure 2**). The aforementioned nanostructures, when modified with fluorescent cores, have applications in the generation of Nanoimages for detection and diagnosis of biological structures and biological events; and controlled release properties [(40).]

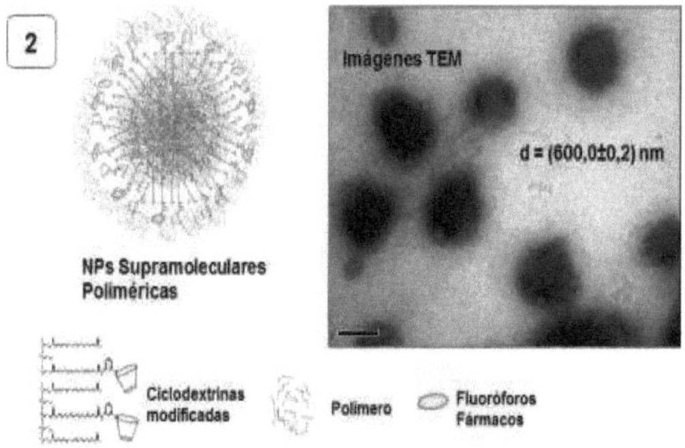

Figure 2. Schematic of Biodegradable Nanoparticles modified with supramolecule systems such as cyclodextrins (CDs), CDs are cyclic alpha-d-glucose oligomers of varying dimensions. Incorporation of Transmission Electron Microscopy (TEM) image of poly-galic-, malic acid nanoparticles. Photos obtained at Laval University, Quebec, Canada. Reprinted with permission from A. Guillermo Bracamonte et al. 2024.

Furthermore, continuing the scaling of dimensions and non-covalent interactions are organised systems such as all those chemical structures where multiple interactions form new architectures of higher dimensions. This is the case of self-assemblies with different molecules and polymers which form micelles, vesicles, DNA and RNA origami, aggregates and lipid nanoparticles, among others. [41] The DNA molecule, deoxyribonucleic acid, is the most important and essential molecule in the development of life. In its sequence is contained all the information of the living being for its development. It has a versatile chemical structure which can be modified and used for the construction of nanostructures, which can be incorporated into living systems with a functionality such as memory devices, biocompatible surfaces, luminescent control, 3D molecular arrays precise control of lengths in the order of nanometres; as well for the study and development of Genomics, Biosensors, Bio-Imaging and diagnostics, Microbiome, bacteria and pathogen generation, personalised Nanomedicine [42].

In this topic related to NanoBioStructures we can mention Origami-type DNA assemblies with which spatially controlled surfaces can be prepared generating 2D and 3D patterns as needed.[43] Due to their structure different fluorescent intercalating agents can be incorporated, as well as the location of Nanoparticles at different distances as needed (**Figure 3**).[(44)]

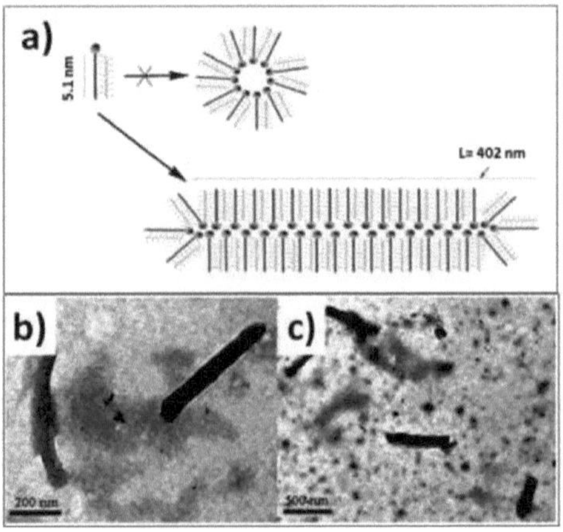

Figure 3. Schematic of origami-like structures based on complementary interactions between DNA strands: a) tubular aggregate model; b) starting from duplexes (double complementary strands); c) in the presence of only 10% of hybridised DNA strands. Reprinted with permission from Denis Boudreau et al. 2024.

In this way Meta-materials, electronic devices, memory generation devices based on photo-electronic transfer by spatial control of organic photo-acceptor molecules and semiconductor nanocomposites, Nanomaterials applied to Nanophotonics by Metal Enhanced Fluorescence (**MEF**),[45] as well as on the basis of their semiconductor properties have been incorporated in OLEDs[(46).]

4.3 Bio-pharmaceuticals and Nutraceuticals

Biopharmacy can be considered as an area of pharmacy which is developed on the basis of bio-products through the incorporation of natural products, extracts, isolated biomolecules and phytochemicals, among others, all from natural sources.[47] In this way, a series of molecules existing in nature can be conceived which are part of the building blocks of biological structures and also participate in the manufacturing process of these structures. Thus, the ingestion of certain nutrients can take on the character of drugs if a healthy and natural anabolic effect is to be induced in the organism. Similarly, a long-term preventative effect is desired.[48]Biopharmaceuticals thus offer a number of products, many of which are widely marketed as nutritional products and are not identified as biopharmaceuticals, but are nonetheless part of a long list of bio-, and/or natural products. In this context, other annexed topics can be mentioned and highlighted, such as functional foods. These are creations of a food in a macroscopic format,real and palpable, which fulfils a specific function. Perhaps the concept can be confusing because all foods have a specific function in our organism, being provided by nature through its natural evolution. Functional foods, on the other hand, are synthetic creations in which the matrix and incorporation of nutrients from different origins are designed in such a way as to direct their function to a specific predefined need.[(49)] In this way, functional foods can fulfil important roles, for example, in neuronal protection at different stages of life according to need, neuronal activation, neuronal nutrition, neuronal constitution and neuronal maintenance. In this case, neurons have been mentioned as important cells which direct the actions, reactions and functions of the entire human body, but can also be applied to other cells and tissues. It is of vital importance to maintain the normal functioning of all cells and organs of the human body; this is done in part by a balanced nutrition in macro- and micronutrients.[(50)] However, functional foods address precisely those situations where the balance is shifted to a state different from the natural optimum and/or is required to maintain that state for a certain period of time.[51] Examples of such situations can be life situations where it is required to demonstrate an organic response to a certain demand such as high performance events. [52] These can range from the conception, gestation and growth of a new being in the mother's womb, to high performance sporting events and extension, known as ultra-performance sporting events.[53] In both examples specific amounts of nutrients, vitamins and minerals are needed which are outside of a need for a particular nutrient, vitamin and mineral.considered as normal.[54] And for its development, the response to the demand will require the organism to unbalance the equilibrium in a sense to comply with the growth event of the new being and/or to cross the sports goal.[55] It is there where functional foods can play important roles for the normal development of life without major consequences after having gone through the nutritional demand. However, it is important to mention that they can not only be applied in extreme events, they can also be used to ensure the controlled intake of nutrients according to need and avoid undesirable situations such as deficiencies and even chronic fatigue events which can easily occur over long periods of time.[56]

Thus, the need for need-based nutrition through functional foods which can be manipulated as processed foods in different formats at the macro-scale has been raised. But this is precisely the control of the macro-scale through the addition of known nutrients and associated effects. This same logic carries over into the infinitely small domain where the micro-, and nanoscale can add different nutrient release strategies as well as interaction mechanisms with living matter. [57] That is, molecules interact with the medium from different media, either dissolved in the cellular medium or after entering the bloodstream. This is simply the concept, which will be further developed in the next sections. And in this context it is important to note that Fundamental Science is elucidating mechanisms of action of nutrients, drugs and chemical agents that are generically recognised as Nutraceuticals.[58]

At this level, molecules and functionalities are the oneswhich direct the action on organisms. Their incorporation is of vital importance, and it is here that the Nanoscale through the design and synthesis of Nanomaterials of different shapes and sizes currently plays a very important role. To get an idea of its dimension, the nanoscale can be compared to an infinitesimal part of the cell surface, which can cover at most, considering a single-celled organism, only 10%.

Continuing with the perspectives of improving and maintaining health by providing nutrients and molecules that contribute in a related aspect, it is important to mention some of the molecules, and biomolecules, among others; such as basic monomeric molecules involved. Thus, different natural polymers known as amino acids, peptides to form proteins, carbohydrates and poly sugars, nucleotides to form genomic material, fatty acids to form molecular assemblies such as membranes, etc. are mentioned. These building blocks are basic and well known, but their all-important wide variety and associated functions should be emphasised. [59]

In addition, we can mention vitamins, antioxidants, neurotransmitters, pro-hormones and hormones, anabolic molecules, stimulants, neurotransmitters, energising molecules, and new molecules that prevent ageing, and increase some desired activity. [60] Similarly, it is important to mention functional biostructures such as the macrobiota, considered as the set of bacteria and micro-organisms found mainly in our digestive system, but also all those found in the whole organism and involved in benefiting its overall functioning. In this sense, there are many studies in progress for the improvement of nutrients and new molecules involved in improving performance and functions [61].

In this context, the development of controlled physical activity should also be considered as part of future treatments, in addition to the intake of nutraceuticals, due to their beneficial effects on health. These can reach the level of modifying membrane compositions and the appearance of marker factors in the blood for the activation of neurogenesis and the development of cognitive processes.[62] Thus, the enzyme phospholipase "glycosylphosphatidylinositol-specifiv phospholipase D1" was discovered in plasma as an important factor in the activation of neurogenesis and stimulation. In addition, the combination of movement and new therapies based on the development of biotechnology manipulating natural components, biostructures, biomolecules, such as beneficial bacteria for health, nutrients, and biocompatible and assimilable synthetic molecules.[63] And from the combination between them the generation of new formulations controlling compositions in search of ergogenic effects and ideally increased, anabolic effects without risking health. In this context, the anabolic state is mentioned as a situation in which a positive construction of biomaterials is generated in the organism, which stimulates cellular and organic activity applied according to need. [64] For example, the manipulation of macrobiota can affect the organism from simple organic functioning to participation in the different cellular mechanisms involved in high-performance sports activities.[65] Thus, sports performances were improved and correlated with the increase of Bacteroides uniformis in the macrobiota. [66] Furthermore, the intake of certain rings of poly-azucars circular with controlled dimensions nominated as cyclodextrins improved performances. It is important to note that these types of compounds, classified as supramolecular systems, have already been tested in controlled release systems for amino acids and proteins in the aerobic and anaerobic sports market.[(67)] This was proposed on the basis of the ability to form host-receptor complexes, which protect the incorporated molecules and release them depending on the interaction from the complex to the outside. If the interaction is stronger at the molecular level, the exchange could occur. Thus, ingestion of such supra-molecules may have ergogenic effects on biomolecule capture in the gut, subsequent release and action. These are potential applications which could be of interest to the market. The aforementioned examples involved molecules and biostructures involved in processes related to the application of a nutrient, nutraceutical or drug; however, other functional structures with intermediate dimensions such as the nanoscale length range can be applied. One can mention silica nanoparticles loaded with curcumin.[68] Curcumin is a chemical produced by a plant called Curcuma longa of the ginger family Zingiberaceae; is sold as a dietary supplement, cosmetic, and food flavouring and colouring agent (**Figure 4**).[69] The mentioned Nanoarchitectures have been studied for their application as new Nanopharmaceuticals with prospects in infectious disease treatment as well as in the food industry. These structures showed mainly improved physico-chemical properties such as solubility, bioavailability, decreased photo degradation, and increased metabolism. Thus, turmeric incorporated into the silica modifies its properties in the process ingestion and metabolism.

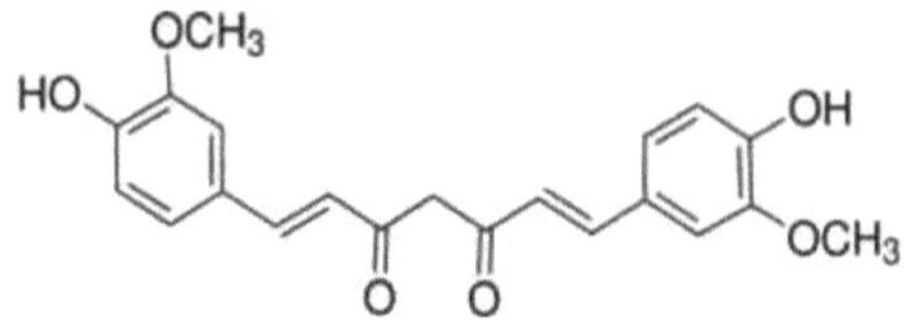

Figure 4. Chemical structure of Curcumin. Reprinted with permission from M. A. Walters et al. 2024.

In other words, a simple organic molecule with important physicochemical and medicinal properties can be used optimally for medicinal and nutritional purposes by modifying its structure by means of a polymeric envelope which protects it and carries it to the site of action. These mentions are only explanatory in order to understand the concept of the use, combination of different compounds with various properties in order to modify and improve properties for their application. Silica is an inorganic polymer which is highly soluble in aqueous and polar media, relatively non-toxic under controlled conditions, and with excellent optical and semiconducting properties which have made it successful in other areas of technology such as devices, chips, photonics, quantum, semiconductors, etc., micro-electronics, circuits, and hard-wares, electronic devices in electronics and computing. Their use has become commercially available from major chemical reagent and instrumentation companies (**Figure. 5**); as well as their precursors for the development of new functional structures which, if used, can be used in the development of new devices and devices. are in the domain of new materials and fundamental science with potential perspectives in different areas of knowledge.

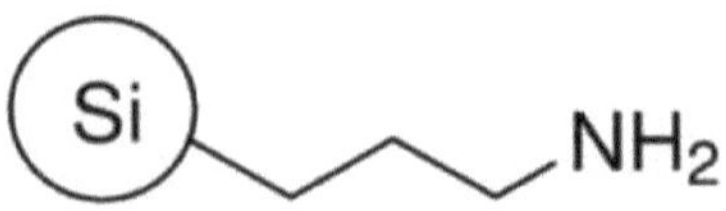

Figure 5. Schematic of Silica Nanoparticles offered by Sigma-Aldrich Company. This represents a sphere with nanoscale dimensions in which various amounts of molecules such as 10-1000 or more can be incorporated depending on their dimensions. Reprinted with permission from Sigma-Aldrich 2024.

The basic nutritional concept of the importance of the macrobiota and the nutrients we ingest needs to be mentioned. In addition, it is important to highlight the different nutritional options available in various sports supplements and pharmaceuticals, both of which must be governed by strict quality controls just like pharmaceuticals. Thus, the domain or area of Nutraceuticals which we intend to showcase and present in this short section is motivated by the discussion of examples such as the above.In this sense, also after having presented nanoscale in next dimensions to nutritionally as well as medicinally active molecules, it can be emphasised more in the visible dimension the microscale for the loading, protection and release of nutrients with superior performances than compared to the intake of free nutrients. Platforms for the release of micro-nutrients on micro-particulate basis [70] This strategy was presented as an efficient way to mitigate micronutrient deficiencies. These encapsulated 11 micronutrients which showed stability to temperature changes during cooking processes, absence of oxidation, and were thus adsorbed by the gut of laboratory rodents. Data from clinical trials demonstrated increased iron loading and release, improved adsorption, increased bioavailability, and thus allowed an increase in the administration of this formulation and production as well. Thus a yield in the order of kg was achieved; and administration from these yields generated similar results as using synthesis on a smaller scale. An 89% higher bioavailability of iron was demonstrated compared to ingestion of the free species.Similarly, the design of novel foods using other strategies based on the use of natural polymeric polymers such as gelatine, sugars, and other biocompatible synthetics are manufactured in processes in which additional external conditions such as high pressure are applied. Thus, polymeric matrices are fused with nutrients, as well as other ingredients within which new nano-, micro- and nutraceuticals can be considered.[71] This technology called "High Pressure Processing (HPP) is in the order of 100-600 MPa; it is applicable for the inactivation of microorganisms, spores, and enzymes in foods and increases the shelf life of the confined functional food. In this way, a review of the subject has been carried out, from which the main topics related to the stated objectives have been presented. The main focus is on the development of nutraceuticals, studying the mechanisms of action of molecules or formulations to positively affect nutrition with a specific interest. [72] However, the use of biological systems such as bacteria, enzymes and mixtures with other components to synergise their actions has also been highlighted. These strategies should be considered for incorporation into cargo systems and functional foods. Where, from molecular level control, to Nano-, Micro-, Micro-scale and larger dimensions allows the design of new formulations and functional foods. In this sense, if we consider other topics related to new perspectives, we should at least mention biotechnology, biomaterials and their incorporation for the use of contact, assimilable, digestible and implantable systems.[73]

And finally, perhaps moving a little further away from the central topic under discussion, we can mention Synthetic Biology, Nanomedicine, and genetic manipulation to advance in a human-controlled way. In this sense, as always, new questions open up with regard to other multidisciplinary topics which can cover ethics, health, sustainability towards a sustainable future.[74]

4.4 Advances in core-shell or core-type cargo nanoarchitectures with functional multi-coatings for non-classical light release.

Several synthesis methods exist to metallic, magnetic, conductive nanoparticles, etc. which are platforms with intrinsic properties to each material that can be chemically modified. Such modifications such as the addition of molecular spacers, stabilising molecules, supramolecular systems, supramolecular coatings, polymeric coatings, etc., can contain drugs for later release. Hybrid materials result in new smart nanomaterials which possess special characteristics such as molecular or cellular recognition, nanoparticle sensing and tracking[75], triggered release by external stimuli,76 biocompatibility and biodegradation.

Regarding the synthesis of Hybrid Nanoparticles, the following has been synthesised Silver core-shell nanoparticles coated with silica, which were modified with various fluorophores using modified organosilanes, and bioconjugated with DNA strands according to the application78 . In addition, nanoparticles have been synthesised with a metallic gold core covered with a biodegradable polymeric coating, to which fluorophores have been added by means of covalent bonds, which have been released at variable times in the order of hours and days, depending on the polymeric composition (**Figure 6**) (Preliminary results shown at the VI Congreso Iberoamericano de Ciencias Farmacéuticas. Pharmaceutical Technology Symposium (COIFFA) 2015. Synthesis of Fluorescent Nanoparticles with a Metallic Core coated with a Biodegradable Polymer applied to the controlled release of drugs. Gontero, D.; Bracamonte, A. G.; Boudreau, D.; Veglia, A. Abstract Book-SAFE 2015 ISSN 2250-4079. B- III-87 - 230)79.

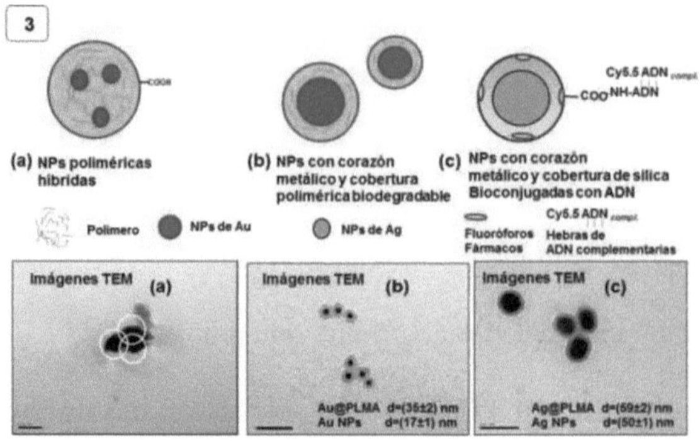

Figure 6. Metal core-type nanostructures with various polymeric coatings: a) incorporation of multi-cores or moulds to form hybrid nanoparticles; b) Nanoparticles with metal cores and biodegradable polymer coatings; c) Nanoparticles with metal cores conjugated with fluorescent silica and DNA strands. Transmission Electron Microscopy (TEM) photos are shown for each of the nanoparticle types. Photos obtained at Laval University, Quebec, Canada. Reprinted with permission from A. Guillermo Bracamonte et al. 2024.

In addition, a proof of concept of a Multifunctional Nanosystem, which incorporates molecular recognition based on Supramolecular interactions, detection based on the Plasmonic effect, such as metal enhanced fluorescence (MEF), control of the luminescent detection signal by a switch based chemical reaction and controlled release of host-receptor complexes, was developed in our laboratory (G. Bracamonte, D. Brouard, M. Lessard-Viger, D. Boudreau, A. Veglia, Microchem. J., 2016,128, 297). fluorescence microscopy we have been able to obtain Nanoimages from Nanoaggregates in the presence of Rhodamine

B concentrations in the order of nM, which corresponds to 40-50 molecules detected per Nanoparticle, with potential applications in unimolecular detection. We are currently working on the development of Ultraluminescent Supramolecular Nanostructures applied to Nanoimaging diagnostics and controlled release systems with applications in Nanomedicine (Preliminary results shown at the Virtual Molecular Diagnostics Symposium 2016. Virtual Event. Labroots, ACS, Fluorescent biodegradable core-shell Nanoparticles applied as platforms for molecular diagnosis by Nanoimaging and drug delivery applications. Gontero, D.; Bracamonte, G.; Boudreau, D.; Veglia, A.)[80].

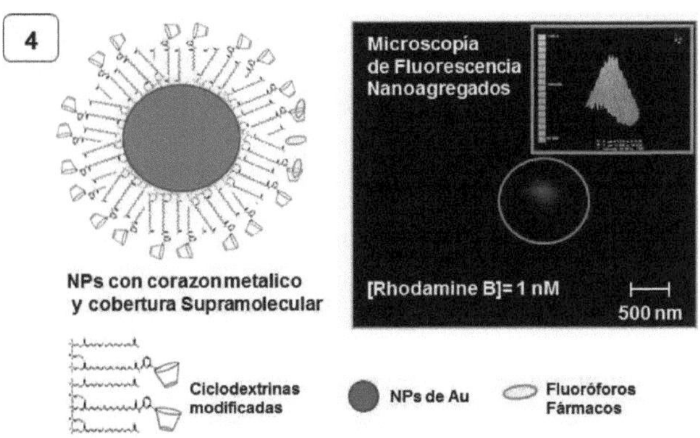

Figure 7. Metal heart-type nanostructures modified with polymeric polyethylene glycol (PEG) chains covalently linked to supramolecular systems such as Cyclodextrins (CDs) Fluorescence Microscopy photos are shown for Nanoassemblies of reduced dimensions. Reprinted with permissions from A. Guillermo Bracamonte et al. 2016.

5. Conclusions and future perspectives

Development from the molecular scale based on the current state of knowledge in the areas focused on in this treatise can be considered for use in the engineering and design of new functional materials for protection and controlled release. However, the approach of different applications produces new situations which cannot be predicted and contribute to fundamental science. This simple fact demonstrates the time required for the transfer of the concept to the desired application. Moreover, the modification of molecules can also make new contributions to the different disciplines of interest presented. In this sense, the need to apply something known, as well as its modification based on a prediction or hypothesis as to its new or potential properties and effects, is highlighted. Similarly, in the development of hierarchical materials from which properties are constructed based on the incorporation of atoms into molecules and structures of greater dimensions. With these general mentions from the point of view of materials and their composition we try to reflect on how different scales can be brought together to have an impact on Macro- and Bio-structures from the point of view of Life Sciences.In this sense, molecules with pharmaceutical activity become important from various points of view such as in the design and synthesis of new molecules and derivatives of natural products. Similarly, Nutraceuticals, as molecules with nutritional activity necessary for; i) the body's constitution, ii) participation in vital processes in the organism,iii) disease prevention, and iv) addressing nutritional deficiencies, v) prevent possible future health problems. Thus, controlled release over varying periods of time is of high interest and impact. In this sense nanostructures can be containers for nutraceutical products which are protected from potential chemical modifications in the course of transport. In addition, micro-scale packaging initiates the scaling up and possible design of functional foods to provide nutrients as needed. In this context, the use of natural compounds for complete biocompatibility and assimilation has been highlighted. However, it is important to highlight the challenges related to the body's acceptance of new formulations. Any intake of new concentrations of nutritional products alters the normal functioning of the body. This is why any concentrated product far removed from the conditions found in nature in a certain way does not respect the natural laws in the process of evolution; but this does not mean that the consumption of concentrations or compositions modified with respect to the natural ones should be avoided as long as they overcome certain basic incompatibilities such as a certain gastric incompatibility. Although these mentions are basic, they are intended to emphasise that although an attempt is made to incorporate a nutrient or nutraceutical product, body's response is always intelligent. This is why the design and strategy of its incorporation should be the same. In other words, it is a challenge open to research and development where certain effects can be designed and predicted, but the real result must be evaluated and validated.On the other hand, with regard to the controlled release of light from the nanoscale, it is important to highlight the design and manufacture of new materials with enhanced or improved properties based on different physical and chemical phenomena. Thus, the synthesis of new hybrid materials applied to collection, energy transfer applied to signal transmission, information encryption, increased photonic efficiency, ultra-sensitive sensors, micro-chips, non-classical light generation with applications in image resolution, and even the development of energy storage on a larger scale for its use is very important in different areas of the Life Sciences. For this reason, the study of energy transfer phenomena at short distances is very important. This involves the physico-chemical study of individual organic molecules, intermolecular interactions, molecules forming part of superstructures, organised systems, Nanostructures, Metamaterials, etc.. In addition, the study of quantum physical and chemical phenomena provides knowledge at the electronic level which is of interest to explain material properties, make predictions and develop applications. This is why synergistic and multidisciplinary work is of vital importance, which is why we are carrying out research and development work on different topics involving luminescent processes, ultrafast luminescent transfer, Metal Enhanced Fluorescence (MEF), Metal Enhanced Fluorescence (MEF), Electron Transfer Electron Reactions (ETR) and Metal Enhanced Fluorescence (MEF), Förster Electron Transfer Reactions (FRET), Chemiluminescence, Electroluminescence and interactions of high energy electromagnetic fields in the near field of Nanostructures with organic molecules, supramolecular systems, organised systems, among others.

ACKNOWLEDGEMENTS

In this section, special thanks are due to all the Grants awarded for the development of Research, Development and Technological Linkage Activities in the different areas discussed in this article. Thus, it is important to mention CONICET, Consejo Nacional de Investigaciones Científicas y Técnicas (National Research Council of Argentine), ANPCyT, Agencia Nacional de Promoción Científica y Tecnológica (National Agency of Scientific and Technology Promotion of Argentine); and especially SECyT (Secretary of Science and Technology from the National University of Cordoba, UNC, Argentina) for awarding the Grant for Young Researchers to the author A. G. B. from INFIQC, FCQ, UNC, Argentina. Professor Denis Boudreau of the "Département de chimie and Centre d'optique, photonique et laser (COPL), Québec, Canada, is thanked for his contribution and collaborative work in progress, as well as for his support in ongoing related grant applications. Thanks are also due to Prof. Valeria Ame of the "Centro de Investigaciones en Bioquímica Clínica e Inmunología (CIBICI), Departamento de Bioquímica Clínica, Facultad de Ciencias, Químicas, UNC", Argentina, for her collaboration on related projects. Special thanks also go to Professor Daniela Quinteros of the Department of Pharmaceutical Sciences, Faculty Chemical Sciences, UNC, together with her working group composed of the D. student Sofia Martinez, and Dr. Cecilia Tettamanti, for collaborative work related to Bioconjugation of

Nanoparticles.Similarly, thanks are due to Professor Burkhardt Koenig of the Department of Organic Chemistry, University of Regensburg, Bavaria, Germany for the possibility of a visit and working mission to his laboratory in 2013, as well as the continuation of his contact until the present day. In addition, thanks are due to Professor Cornelia BohneUniversity of Victoria, British Columbia, Canada, for the opportunity for Postdoctoral experience related to the Design, Synthesis and Dynamics of Supramolecular Systems. As well as for the work carried out during this stay with the Petrochemical Industrial Sector "Oil Alberta", Alberta, Canada. Similar thanks are due to Professor Nita Sahai in her laboratory at the Department of Polymer Science, Good Year Polymer Research Center, and NASA Astrobiology Institute, University of Akron, Ohio, USA in 2013-2014. Similarly, thanks are due to Prof. Ahmed E. Shalan from "BCMaterials, Basque Center for Materials, Applications and Nanostructures, also associated with Central Metallurgical Research and Development Institute (CMRDI)", Egypt; for his participation together with his student Esraa S. A. Serea in recently published papers in the context of a collaborative work. In addition, special thanks to Prof. and Researcher E. Garcia - Quismondo of the Electrochemical Processes Unit, IMDEA Energy, Mostoles, Madrid, Spain.Finally, special thanks to the Education Commission "Des Decouvertes", Quebec; and "British Columbia Education system", BC, Canada for their contributions in International Education and Multiculturalism. And in this context to the participation of the Argentine Education System in the different stages involved in these activities.Finally, thanks are due to all for the recent technology-based research and development venture "Entrepreneurship-Start-Up: Bio-highlighting solutions" founded by the author A. G. B. et al., since its inception with Awarded Prix Ideas Challenge 2010, Entrepreneuriat.

ULaval-Université Laval,Laval University, Quebec, Canada.

(https://www.eul.ulaval.ca/).

BIBLIOGRAPHY

1 S. Bhattacharya, K. M. Alkharf , R. Janardhanan,D. Mukhopadhyay, Nanomedicine: pharmacological perspectives, Nanotechnol Rev, 1 (2012) 235-253

2 C. M. Hussain, Editor-in-Chief, Handbook of Functionalized Nanomaterials for Industrial Applications. A volume in Micro and Nano Technologies , 2020Copyright © 2020 Elsevier Inc. 1-520. ISBN 978-0-12-821381-0

3 A. Guillermo Bracamonte*, Luna R. Gomez Palacios, Book entitled: "Functional Nano-Biostructures. Study of Dynamics and Luminescent properties based on MEF with perspectives within Micro-devices and Bio-lasers", Editorial Académica Española: OmniScriptum GmbH & Co. KG, Heinrich-Böcking-Str. 6- 8, 66121, Saarbrücken (Germany). Directors: Marta Lusena and Dr. Wolfgang Philipp Müller. ISBN 9786202102360. (2023) 1-76. https://www.eae-publishing.com/

4 M. Rioux, D. Gontero, A. V. Veglia, A. Guillermo Bracamonte*, D. Boudreau*, Synthesis of Ultraluminiscent gold core-shell Nanoparticles as NanoImaging Platforms for Biosensing applications based on Metal enhanced fluorescence, RSC Adv., 7 (2017) 10252-10258.

5 C. Wang et al., Advanced Nanotechnology Leading the Way to Multimodal Imaging-Guided Precision Surgical Therapy, 31, 49, 6 (2019) 190432.

6 Michael A. Pinkert, Lonie R. Salkowski, Patricia J. Keely, Timothy J. Hall,

Walter F. Block, Kevin W. Eliceiri, Review of quantitative multiscale imaging of breast cancer, J. Med. Imag. 5, 1, 010901 (2018) 1-20.

7 Daniel Volz, , Review of organic light-emitting diodes with thermally activated delayed fluorescence emitters for energy-efficient sustainable light sources and displays, J. Photon. Energy,6, 2, 020901 (2016) 1-13.

8 A.Guillermo Bracamonte, Book entitled: "Frontiers in Nano- and Micro-device Design for Applied Nanophotonics, Biophotonics and Nanomedicine", Chapters 1-15, authored Book, Bentham Science Publishers, ISBN: 978-1-68108-857-0 (Print); 978-1-68108-856-3 (Online)© 2021, Bentham Science Publishers (UAE) (2021) 1-200.
https://benthambooks.com/book/9781681088563/
DOI:10.2174/97816810885631210101

9 L. R. Gomez Palacios, A. G. Bracamonte*, Development of Nano-, Microdevices for the next generation of Biotechnology, Wearables and miniaturized Instrumentation, RSC Adv., 12 (2022) 12806-12822

10 T. Pradeep, NANO. The Essentials. Understanding nanoscience and nanotechnology, Mc Graw Hill, 2007.

11 Kimling J., Maier M., Okenve B., Kotaidis V., Ballot H., Pletch A. Turkevich method for gold nanoparticle synthesis revisited. J. Phys. Chem. B (2006) 110: 15700-15707.

12 C. Graf,D. L. J. Vossen, A. Imhof, A. van Blaaderen, A General Method to Coat Colloidal Particles with Silica, Langmuir 19 (2003) 6693-6700.

13 Murakami H. et al. Preparation of poly(dl-lactide-co-glycolide) nanoparticles by modified spontaneous emulsification solvent diffusion method. Int. J. of Pharm. (1999) 187:143-152.

14 A. G. Bracamonte, G. Miñambres, N. Pacioni, D. Boudreau, A. V. Veglia, Supramolecular Systems applied to the synthesis of new Nanostructures, by INFIQC, Departamento de Química Orgánica, Facultad de Ciencias Químicas, Universidad Nacional de Córdoba. Ciudad Universitaria, 5000 Córdoba, Argentina, and Departement de chimie and Centre d'optique, photonique et laser (COPL), Université Laval, Québec (QC), Canada, G1V 0A6. Digital logbook Journal. New Materials, Vol. 2, 6th Ed., Faculty of Chem. Sc. (UNC), (2015) ISNN: 2344-9144.

https://revistas.unc.edu.ar/index.php/Bitacora/issue/archive

15 C. N. R. Rao, A. Müller, A. K. Cheetham, The Chemistry of Nanomaterials, WILEY-VCH Verlag GmbH & Co. KGaA, 2004.

16 Goldberg M, Langer R, Jia X. Nanostructured materials for applications in drug delivery and tissue engineering. J Biomat Sci. (2007) 18:241-68.

17 Panyan J, Labhasetwar V. Biodegradable nanoparticles for drug and gene delivery to cells and tissue. Adv Drug Deliv Rev (2003) 55:329-347.

18 Moddaresi M, Brown MB, Zhao Y, Tamburic S, Jones SA. The role of vehicle nanoparticle interactions in topical drug delivery. Intern J Pharmac. (2010) 400 (1-2):176-182.

19 Bracamonte A. Guillermo, Veglia Alicia V. Cyclodextrins nanocavities

effects on basic and acid fluorescence quenching of hydroxy-indoles. Journal of Photochemistry and Photobiology A: Chemistry (2013) 261:20- 25.

20 M. R. Begley , D. S. Gianola , T. R. Ray, Bridging functional nanocomposites to robust macroscale devices, Science, 364, 6447, eaav4299 (2019) 1-10.

21 A. Guillermo Bracamonte*, D. Brouard, M. Lessard-Viger, D. Boudreau*, A.V. Veglia, Nano-supramolecular complex synthesis: switch on/off enhanced fluorescence control and molecular release using a simple chemistry reaction, Microchemical Journal, 128 (2016) 297-304.

22 A. Guillermo Bracamonte*, Gold nanoparticles chemical surface modifications as versatile Nanoplatform strategy for fundamental Research towards Nanotechnology and further applications, Nanoscience and Nanotechnology: Open Access; MedDocs Publishers LLC, 1, 1, 1, 1004. (2022) 1-6.

23 Luna R. Gomez Palacios, Sofia Martinez, Cecilia Tettamanti, Daniela Quinteros, A. Guillermo Bracamonte*, Nano-chemistry and Bio-conjugation with perspectives on the design of Nano-Immune platforms, vaccines and new combinatorial treatments, J Vaccines Immunol, PeerTechz Editorial; 7, 1 (2021) 049-056

24 Alexis F., Pridgen E.M., Langer R., Farokhzad O.C., Nanoparticle technologies for cancer therapy, Handb. Exp. Pharmacol. (2010) 1:55-86.

25 Nie S., Xing Y., Kim G.J., Simons J.W., Nanotechnology applications in

cancer, Annu. Rev. Biomed. Eng. (2007) 9:257-288.

26 Ayelen Inda, Sofia Mickaela Martinez, Cecilia Tettamanti, Carolina Bessone, Daniela Quinteros, A. Guillermo Bracamonte, Chapter 7 Book entitled: "Nano-engineering Multifunctional Organized Systems highlighting Hybrid micelles, vesicles and Lipidic aggregates towards higher sized structures for theranostics perspectives", to be incorporated in the Elsevier Science and Technology reference book on "Theranostics nanomaterials in drug delivery, Edited by Prof. Prashant Kesharwani (India, UK), and Timothy Bennet, Ed. Prashant Kesharwani (India, UK), and Timothy Bennet, Ed. Project Manager of Life Sciences Elsevier, ISBN: 978-0-443-22044-9, Elsevier (Netherlands), Published On Line by Oct. 16, 2024 (2025) 111-132. DOI: https://doi.org/10.1016/B978-0-443-22044-9.00020-6 © 2025 Elsevier Inc.USA.

27 S. Bhattacharya, K. M. Alkharfy, R. Janardhanan, D. Mukhopadhyay, Nanomedicine: pharmacological perspectives, Nanotechnol Rev, by Walter de Gruyter, Berlin, Boston 1 (2012): 235-253.

28 Y. Pathak, D. Thassau, Editors, Drug Delivery Nanoparticles, formulations and characterization. Drugs and the Pharmaceutical Sciences, Vol. 191, Inform Health Care, New York (2009). ISBN-13: 978-1-4200-7804-6 (hardcover : alk. paper) ISBN-10: 1-4200-7804-6 (hardcover : alk. paper)

29 V. Mercadante, E. Scarpa, V. De Matteis, L. Rizzello, A. Poma , Engineering Polymeric Nanosystems against Oral Diseases, Molecules, 26, 2229 (2021) 1-35.

30 C. Tomuleasa, C. Braicu, A. Irimie, L. Craciun, Ioana B.-Neagoe Nanopharmacology in translational hematology and oncology, International Journal of Nanomedicine, 9 (2014) 3465-3479.

31 Q. Ni et al., A bi-adjuvant nanovaccine that potentiates immunogenicity of neoantigen for combination immunotherapy of colorectal cancer, Sci. Adv. 6 : eaaw6071 (2020) 1-12.

32 A. Kumari, S. Kumar Yadav, S. C. Yadav, Biodegradable polymeric nanoparticles based drug delivery systems, Colloids and Surfaces B: Biointerfaces 75 (2010) 1-18.

33 G. Li, M. Zhao ,F. Xu, B. Yang, X. Li, X. Meng, L. Teng , F. Sun, Y. Li, Molecules, 25, 5023 (2020) 1-18,.

34 J. Ding, Y. Li, Editors, Smart Functional Polymers, Molecules, MDPI (2022) 1-308. ISBN 978-3-03921-590-4 (Pbk)35

36 Daniela Gontero, Mathieu Lessard-Viger, Danny Brouard, A. Guillermo Bracamonte, Denis Boudreau, Alicia V. Veglia, Smart Multifunctional Nanoparticles design as Sensors and Drug delivery systems based on Supramolecular chemistry, Microchemical Journal, 130 (2017) 316-328.

37 N. L. Pacioni, A. G. Bracamonte, A. V. Veglia, Comparative effects of cyclodextrin nanocavities versus organics solvents on the fluorescence of indole and carbamate compounds, J. Photochem. Photobiol. A: Chem. 198 (2008) 179-185.

38 Y. Jiang, Y. Liang, H. Zhang,W. Zhang, S. Tu, Preparation and

biocompatibility of grafted functional □-cyclodextrin copolymers from the surface of PET films, Materials Science and Engineering C 41 (2014) 1-7.

39 Davis M. E., Zuckerman J. E., Choi C. H. J. Evidence of RNAi in humans from systemically administered siRNA via targeted nanoparticles. Nature (2010) 464 :1067-1071.

40 M. Friederike Schulte, Steffen Bochenek, Monia Brugnoni, Andrea Scotti, AhmedMourran,Walter Richtering, Stiffness Tomography of Ultra-Soft Nanogels by Atomic Force Microscopy, Angew. Chem. Int. Ed., 59 (2020) 2-9.

41 Eleonora Trovato, Valentina Di Felice, Rosario Barone, Extracellular Vesicles: Delivery Vehicles of Myokines, Frontiers in Physiology, 10, 522 (2019) 1-13.

42 X. Tian, S. Angioletti-Uberti, G. Battaglia, On the design of precision nanomedicines, Sci. Adv.;6: eaat0919 (2020) 1-11.

43 Y. C. Hung , D. M. Bauer , I. Ahmed, L. Fruk , DNA from natural sources in design of functional devices, Methods, 67, 2, (2014) 105-115.

44 Kim Dore, R. Neagu-Plesu, Mario Leclerc, Denis Boudreau, Anna M. Ritcey,, Characterization of Superlighting Polymer-DNA Aggregates: A Fluorescence and Light Scattering Study, Langmuir 2007, 23, 258-264.

45 J. Asselin, M. L.Viger, D. Boudreau, Review: Article Metal-Enhanced Fluorescence and FRET in Multilayer Core-Shell Nanoparticles, Advances in Chemistry, Hindawi Publishing Corporation, 812313, (2014) 1-16.

46 M. Lunza, D. de Boera, G. Lozano, S. R K Rodriguez, J. Gomez Rivas,

M. A Verschuurena, Plasmonic LED device, Proc. of SPIE, 9127, 91270N (2014) 1-6.

47 G. Walsh, Biopharmaceutical benchmarks, Nature Biotechnology, 18 (2000). 831-833.

48 Vivek Puri et al., A Comprehensive Review on Nutraceuticals: Therapy Support and Formulation Challenges, Nutrients, 14 (2022) 4637.

49 N. J. Temple, A rational definition for functional foods: A perspective, Front. Nutr., Sec. Nutrition Methodology, 9 (2022)

50 R. Doggui,; H. Al-Jawaldeh,; J. El Ati,; R. Barham, L. Nasreddine; N. Alqaoud,; H. Aguenaou, L. El Ammari, J. Jabbour, A. Al-Jawaldeh, Meta-Analysis and Systematic Review of Micro- and Macro-Nutrient Intakes and Trajectories of Macro-Nutrient Supply in the Eastern Mediterranean Region. Nutrients, 13, 1515. H (2021) 1-19.

51 Clare M. Hasler, Functional Foods: Benefits, Concerns and Challenges-A Position Paper from the American Council on Science and Health, American Society for Nutritional Sciences, 0022-3166/02 (2002) 372-381.

52 C. Thurber, L. R. Dugas, C. Ocobock, B. Carlson, J. R. Speakman, H. Pontzer, Extreme events reveal an alimentary limit on sustained maximal human energy expenditure, Sci. Adv. 2019;5: eaaw0341(2019) 1-8.

53 K. A. Hammond, M. Konarzewski, R. M. Torres, J. Diamond, Metabolic ceilings under a combination of peak energy demands. Physiol. Zool. 67, 1479-1506 (1994) 1-10.

54 G. Siuzdak, Metabolic adaptation to calorie restriction, Sci Signal. 8;13, 648:eabb2490 (2020) 1-10.

55 K. L Beck, J. S Thomson, R. J Swift, P. R von Hurst , Role of nutrition in performance enhancement and postexercise recovery, Open Access J Sports Med, 11, 6 (2015) 259-267.

56 Maria Ester la Torre et al., The Potential Role of Nutrition in Overtraining. Syndrome: A Narrative Review, Nutrients, 15, 23:4916 (2023) 1-18.

57 1st Edition, Handbook of Nanotechnology in Nutraceuticals, Edited By S. Ahmed, T. Bhattacharya, Annu, A. Ali, Taylor-Francis Goupe, Routeledge, CRC Press (2024) 1-484. ISBN 9781032155678

58 M. Bernela, P. Kaur, M. Ahuja, R. Thakur, Nano-based Delivery System for Nutraceuticals: The Potential Future. In: Gahlawat, S., Duhan, J., Salar, R., Siwach, P., Kumar, S., Kaur, P. (eds) Advances in Animal Biotechnology and its Applications. Springer, Singapore. https://doi.org/10.1007/978-981-10-4702-2_7

59 A. E, Khalil, Mainardes RM, Bovine serum albumin nanoparticles containing quercetin: characterization and antioxidant activity. J Nanosci Nanotechnol, 16 (2016) 1346-1353

60 T. N. Ziegenfuss, J. M. Berardi, L. M. Lowery, J. Antonio, Effects of Prohormone Supplementation in Humans: A Review, Canadian Journal of Applied Physiology, 27, 6 (2002) 1-10.

61 P.A. Vlaicu, A.E. Untea, I. Varzaru, M. Saracila, A. G. Oancea, Designing Nutrition for Health- Incorporating Dietary By-Products into Poultry Feeds to

Create Functional Foods with Insights into Health Benefits, Risks, Bioactive Compounds, Food Component Functionality and Safety Regulations. Foods 12, 4001. H (2023) 1-36.

62 A. M. Horowitz , X. Fan , G. Bieri , L. K. Smith , I. Sanchez-Diaz , A. B. Schroer , G. Gontier , K. B. Casaletto , J. H. Kramer , Katherine E. Williams , S. A. Villeda, Blood factors transfer beneficial effects of exercise on neurogenesis and cognition to the aged brain, Scienceebl (2020) 144-149.

63 N. Hadadi, V. Berweiler, H. Wang, M. Trajkovski, Intestinal microbiota as a route for micronutrient bioavailability, Current Opinion in Endocrine and Metabolic Research, 20, 100285 (2021) 1-23.

64 J. Trommele, The anabolic response to protein ingestion during recovery from exercise has no upper limit in magnitude and duration in vivo in humans, Cell Reports Medicine,4, 101324, 19 (2023) 1-22.

65 S. Moochhala et al., Intestinal Microbiota Interventions to Enhance Athletic Performance-A Review, Int. J. Mol. Sci. 25, 18 (2024) 1-20.

66 H. Morita, C. Kano, C. Ishii, N. Kagata, T. Ishikawa, A. Hirayama, Y. Uchiyama, S. Hara, T. Nakamura, S. Fukuda, Bacteroides uniformis and its preferred substrate, a-cyclodextrin, enhance endurance exercise performance in mice and human males, Sci. Adv. 9, eadd2120 (2023)1-17.

67 P.Bucur, I. Fülöp, E, Sipos, E., Insulin Complexation with Cyclodextrins-A Molecular, Modeling Approach. Molecules, 27, 465 (2022)1-20.

68 S. Maleki Dizaj, S. Sharifi, F. Tavakoli, Y. Hussain , H. Forouhandeh, S.

Mahdi Hosseiniyan Khatibi , M. Y. Memar , M. Yekani, H. Khan, K. Wen Goh, L. Chiau Ming, Curcumin-Loaded Silica Nanoparticles: Applications in Infectious Disease and Food Industry, Nanomaterials, 12 (2022) 2848 1-16

69 KM Nelson, JL Dahlin, J Bisson, J Graham, GF Pauli, MA Walters, The Essential Medicinal Chemistry of Curcumin, Journal of Medicinal Chemistry. 60, 5 (2017) 1620-1637

70 A.C. Anselmo et al., A. Jaklenec, A heat-stable microparticle platform for oral micronutrient delivery, Science Translational Medicine, 11, 518, eaaw3680 (2019) 1-10. DOI: 10.1126/scitranslmed.aaw3680

71 A.Kaushal Balakrishna, M. Abdul Wazed, M. Farid,A Review on the effect of high pressure processing (HPP) on Gelatinization and Infusion of Nutrients, Molecules 2020, 25, 2369 1-19.

72 L. Liu, W. Zhao, Q. Ma, et al. Functional nano-systems for trans-dermal drug delivery and skin therapy. Nanoscale Adv. 5 (2023) 1527-1558.

73 L. Fan, J. Huang, S. Ma, Recent advances in delivery of transdermal nutrients: A review, 33, 1, e14966 (2024) 1-14.

74 J. Chen, Synthetic biology for future food: Research progress and future directions, Future Foods, 3, 100025 (2021) 1-12.

75 Y-Ammar A., Sierra D., Mérola F., Hildebrandt N., Le Guével X., Self-Assembled Gold Nanoclusters for Bright Fluorescence Imaging and Enhanced Drug Delivery. ACS Nano (2016) 10, 2: 2591-2599.

76 Bracamonte A. Guillermo, Brouard D., Lessard-Viger M., Boudreau D.,

Veglia Alicia V. Nano-supramolecular complex synthesis: switch on/off enhanced fluorescence control an molecular release using a simple chemistry reaction. Microchemical Journal (2016) 128:297-304.

77 Brouard D., Ratelle O., Bracamonte A. Guillermo, St-Louis M., Boudreau D. Direct molecular detection of SRY gene from unamplified genomic DNA by metal enhanced fluorescence and FRET. Analytical Methods (2013) 5:6896 - 6899.

78 Brouard D.; Lessard Viger M.; Bracamonte A. G.; Boudreau D. Nanoplasmonic sensing of nucleic acids using fluorescent core-shell nanoparticles and a cationic fluorescent polymer. ACS Nano (2011) 5:1888-1896.

79 Synthesis of Fluorescent Metal Core Nanoparticles coated with a Biodegradable Polymer applied to the controlled release of Drugs. Gontero, D. [2]; Bracamonte, A. G. [1]; Boudreau, D. [2]; Veglia, A. [1]. [1] Instituto de Investigaciones en Físicoquímica de Córdoba (INFIQC), Departamento de Química Orgánica, Facultad de Ciencias Químicas, Universidad Nacional de Córdoba. Ciudad Universitaria, 5000 Córdoba, Argentina [2] Dep. de chimie et Centre d'optique, photonique et laser (COPL), Université Laval, Québec, Canada. VI Iberoamerican Congress of Pharmaceutical Sciences. Pharmaceutical Technology Symposium (COIFFA), 4-6 November 2015, Córdoba, Córdoba, Argentina. Book of Abstracts-SAFE 2015 ISSN 2250-4079 (B- III-87 - 230).

80 Fluorescent biodegradable core-shell nanoparticles applied as platforms for molecular diagnosis by nanoimaging and drug delivery applications. Gontero, D. [2]; Bracamonte, G. [1]; Boudreau, D. [2]; Veglia, A [1]. INFIQC, Dept. of Organic Chemistry, Faculty of Chemical Sciences, National University of Cordoba. Dep. de chimie et Centre d'optique, photonique et laser (COPL), Université Laval, Québec, Canada. Molecular Diagnostics 2016. Virtual Event. Labroots, ACS 6-8 April 2016.

MIX
Papier aus verantwortungsvollen Quellen
Paper from responsible sources
FSC® C105338

Printed by Books on Demand GmbH, Norderstedt / Germany